The Places We Live

Contents

Natural Wonders

The United States of America has an area of 9,161,922 square kilometers (3,537,438 miles). Its has all kinds of **landforms** and features of Earth's surface.

There are coasts, mountains, flat open spaces, deserts, lakes, rivers, and forests. Do you live near one of these landforms or features?

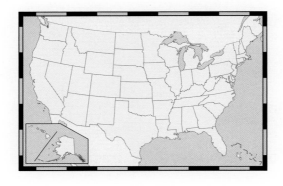

What is the relationship between kilometers and miles? Which number is bigger? Smaller?

Near the Coast

The coast of Florida is about 1,350 miles long. Akira's home in Florida is near the Gulf of Mexico. He likes to swim and surf.

Akira and his friend walk 15 minutes to get to the beach. They live only 1 mile from the beach.

TALK ABOUT IT

Explain how you could find the total amount of time that Akira walks to and from the beach in a week.

In the Mountains

Olivia's home is in Colorado. She lives near the Rocky Mountains. Colorado is more than 2,000 feet above sea level.

Olivia's family hikes on a trail that is 6 kilometers (4 miles) long. They have seen deer, golden eagles, and red foxes on the trail.

TALK ABOUT IT

What do you think *2,000 feet above sea level* means?

Rolling Hills

Kwan lives in North Carolina. Her family has a large farm. They grow and sell fruit.

Kwan and her grandfather **harvest** peaches. They harvested 12 pounds (5.4 kilograms) of peaches today.

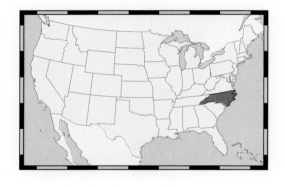

In the Desert

Edgar explores in the desert near his home in Texas. He **examines** each animal he finds. He draws pictures of the animals. Edgar writes the length and color of the animals he sees.

Desert Log Book

Date	Animal name	length	Color
April 6	desert tortoise	15 inches	brown gray
April 7	hairy desert Scorpion	about 4 inches	yellow
April 9	Queen butterfly	3 inches	orange, black + white
April 10	desert toad	5 inches	brown
April 12	tarantula	$6\frac{1}{2}$ inches	black
April 18	desert iguana	10 inches	tan, green + brown

Desert toad →

TALK ABOUT IT

Describe how Edgar could measure the animals.

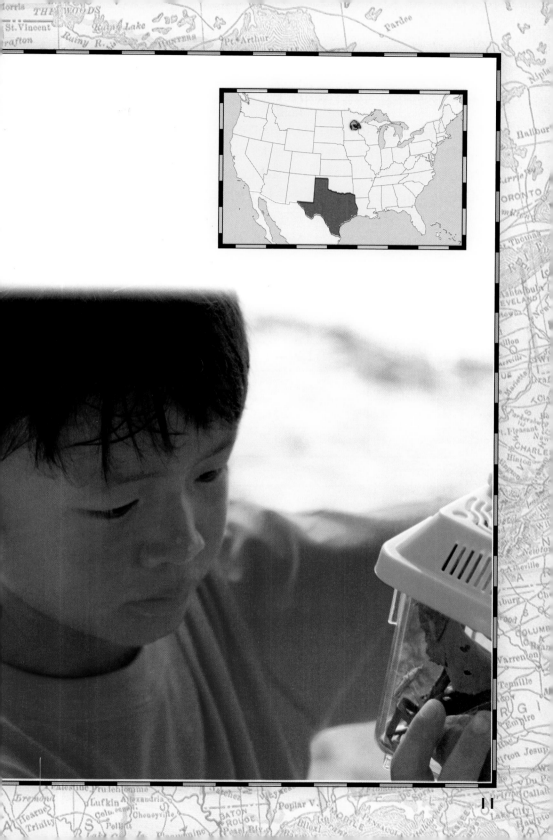

Lake Country

Toya and her family live near the Great Lakes in Michigan. Her family likes to sail. It takes Toya's family about 20 minutes to get to Lake Superior. They live 16 kilometers (10 miles) from the lake.

Water Area of the Great Lakes	
Lake	**Area** (square miles)
Superior	31,700
Michigan	22,300
Huron	23,000
Erie	9,910
Ontario	7,340

Did You Know?

The Great Lakes are made up of five lakes. They are the largest group of freshwater lakes in the world!

Under the Trees

David lives in northern California. He and his father walk in a forest of redwood trees. They like to measure the **circumference** of the tree trunks. They measure the distance around the base of a different tree each day.

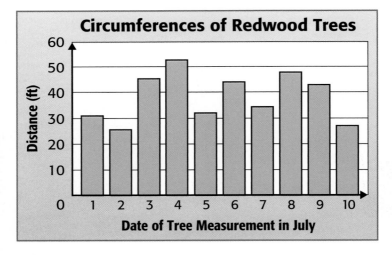

David writes the tree measurements on this bar graph. Each day he adds a new bar. The bar shows the circumference of the tree measured that day.

TALK ABOUT IT

Describe two ways that David could measure the circumference of a tree.

Glossary

circumference
> Measurement of the distance around something circular. *(page 14)*

examine
> To carefully look at and study something. *(page 10)*

harvest
> To pick and gather up a crop that is ripe. *(page 8)*

landforms
> Natural features of the land, such as mountains, lakes, valleys, plains, or canyons. *(page 2)*